COLD KILLS
WARMTH GIVES LIFE

BRUCE WILLIAMS

The permission address is
Permissions
PO Box 3938
Gillette, WY 82717

ISBN Paperback: 978-1-7358771-0-5
ISBN Electronic: 978-1-7358771-1-2

Library of Congress Control Number: 2020920001

Printed in the United States of America.

Dedication

Dedicated to the scientific process and the search for truth.

Contents

Introduction

Welcome to the "560" book. I hope you will find this information both informative and as a guide to helping all mankind.

This is in no way an attempt to deny that carbon dioxide is a greenhouse gas and warms up when exposed to infrared radiation. Neither is it an attempt to deny that increasing carbon dioxide in the atmosphere changes the climate; or to argue that our climate is not changing. Denying premises backed by experimental results is not only a waste of time but counterproductive. Alternatively, questioning how known laws are applied is the foundation of all sciences.

I have written this not as a paper for peer review but rather as a book for those who want to know a little more about the technical information without being swamped by it.

In the section **Sources & Notes** you will find my references which are books, peer-reviewed published articles, or information from government and educational websites.

The objective of this book is to show what the errors are that explain the four conditions below, why they are dangerous to the point of possible elimination of life on earth, and what can be done about the situation we are in.

Unfortunately, politicians and others chose the most dangerous path possible. This path is dangerous because:

1. Cold kills: warmth gives life.
2. Carbon dioxide enhances plant and animal life on earth.
3. The earth is headed into uncharted astronomical waters.
4. We may not be able to stop the warming because the primary driver may not be carbon dioxide.

As we are all aware, there has been a bit of paranoia around the planet for the last 40 years. This irrational fear was labeled global warming and like all fear campaigns it has some basis in truth but is mostly a psychological war to encourage people to do something. The idea is to get people to use fewer fossil fuels and instead use "renewable" power sources.

There is not one single factor driving this irrational fear, but all the factors are associated with money. Whether it is a desire to extend political power and influence, gain funding for research, ego massaging, gaining more manufacturing diversity, growing the economy, or whatever, there are many people who see it as an opportunity and jump on the band wagon. What you must understand is there are a lot of people with either a monetary or ego stake in what the movement is doing and, right or wrong, it is not going away.

Chapter 1

The Pascal's Wager Fallacy

Seventeenth century philosopher, theologian, mathematician, and physicist, Blaise Pascal, theorized that humans should be willing to bet their lives on whether or not God existed[1]. Pascals wager had one major flaw, if you select the wrong God then the true God would punish you. In other words limiting yourself to only 2 options when there are multiple options is flawed. One of the earliest environmental arguments was applying Pascal's Wager to global warming. In this case the logic was:

1. Continue burning fossil fuels - which leads to global warming - and you cause a major catastrophe.
2. Quit burning fossil fuels - which leads to global warming - and you have prevented a major catastrophe.

3. Continue burning fossil fuels - which does not lead to global warming - and no harm is done.
4. Quit burning fossil fuels - which does not lead to global warming - and no harm is done.

Obviously, this would indicate that to stop burning fossil fuels leads to two good outcomes. Continuing to burn fossil fuels leads to one bad and one good outcome. The one bad outcome plus the other three good outcomes shows that the obvious path is to stop burning fossil fuels - at least in this simple picture.

You will notice that no one asked, "What happens if we stop burning fossil fuels and it resulted in harm? Would it be more harmful than more carbon dioxide in the air?"

Of course, everyone said, "Pshaw - there's nothing in our understanding of the laws of physics that can do that. You are just theorizing and have nothing to back up your story. The harm is in warming up the earth and melting all the ice and making more storms and hurricanes and all that. This is true because the scientists told us so."

The serious scientists never said such a thing. They said - I don't know - could be, could not be, we don't know for sure.

But the politician's response was, "Leaders sometimes need to act quickly with incomplete information." So, they started handing out money and organizing people to grab more power for themselves and encouraging other politicians to fight this global warming epidemic. There were going to be carbon credits to buy and sell, limits on this and that, requirements to buy certain amounts of new technology, etc.

And all the world's great knights in shining armor ran off to slay the "Dragon of Global Warming" and never looked back.

But the nerdy people did look back, as they often do, and began to ask questions. Questions such as, "Why aren't there more hurricanes, or at least more powerful hurricanes? Why wasn't the ice melting as much as it should? Why didn't the atmosphere warm up like the climate models (equations) said it should? Why did all the climate models (equations) raise the temperature of the earth more then what was actually occurring? Why wasn't the ocean rising like it should have? " And many more.

Chapter 2

All The Fundamental Laws Of Physics

What do equations have to do with climate models? I'm sure you are familiar with the famous equation $E = mc^2$. This equation mathematically describes something that was noticed about any subatomic particle when it's mass changes. And it is accepted in science because no matter what, if the sub atomic particles mass changes the energy given off if it loses mass or the energy it absorbs if it gains mass will be exactly the value calculated in the equation. EVERY time. We know it will not just be the majority of the time, but every time because this equation is used in the development of stellar astrophysics, nuclear bombs, nuclear reactors, particle accelerators, x-ray machines, and many others and is always correct.

To calculate what the climate is going to do, you need mathematical equations to describe physical properties. But in the case of the climate you need a bunch of them, a whole bunch.

So, these large number of equations are put into computers, resulting in a single calculation that is fed into a few other equations which feed other equations, and on and on. This is called the model. Some of the equations then loop back to the start and that is added to the beginning as the new start. The whole process is repeated over and over until a result is obtained and finally graphed, put into a spread sheet, a database, or printed.

When combined, these multiple equations must produce something like the $E = mc^2$ equation. A result that is right all the time. Not just one or two points matching up with reality over a limited period of time, but the vast majority of the results have to match. With this many equations you'll never get a perfect match but you should stay within a few percentage points of what is really happening. Every mass to energy conversion that has been measured does not exactly match the equation because of the inaccuracy of our measuring instruments, but the equation answer is always within the error range of the measuring instrument(s).

And therein lies the problem. All the settled science can't produce anything that agrees within the accuracy of our instruments. This mismatch is verified by the lack of no more than normal hurricanes and tornadoes, the atmosphere not warming up as predicted, and many more physical and climatic facts.

So, what do we know that can guide us in deciding what is best for mankind? Obviously, the models (equations) aren't good enough, so how about we look at reality?

How can I validly compare the present situation to past values? After all, the rise in carbon dioxide is not caused by nature, it is mankind that is introducing the carbon dioxide. One must realize that what is introducing the carbon dioxide has nothing to do with the effect of introducing carbon dioxide into the environment as long as the only factor changing is the quantity of carbon dioxide in the atmosphere. Mankind introducing the carbon dioxide is no different than any of the 17 major basaltic flood plains (traps) that have occurred naturally. And of these 17, only three (in combination with other factors) did any harm to the earth. Thus, it should be obvious that the results of these actual events is what the models should be able to reproduce. But they cannot and as such the models are wrong. The actual response is right because it is literally what happened.

Reality or Model?

The model for growing a field of wheat says you will lose X bushels per acre every year due to reduced nitrogen. But the reality is farmer Jones' field has continued to produce the same number of bushels of wheat every year. One would think the people who wrote the model would say their model is not complete and needs more data rather than "adjusting" farmer Jones' production numbers to match the model output.

We are in the habit of adjusting data to make the real world match the model rather than the other way around. I consider the comparison of what has happened in the past a much more reliable indicator of what will happen in the future rather than relying on a model which cannot reproduce the physical world.

This is the point that is missed in all the debates on global warming. And this is a decision each and every person who chooses what to do must make. You must choose whether to rely on what has actually happened or on a human generated model that does not match reality.

When people say something is going to happen, no one has asked if their viewpoint is based on the physical world's response (reality) or on the output from a computer programmed by humans?

Chapter 3

The Culprits – Temperature And CO2

First, let's look at the "culprits" everyone is so afraid of: carbon dioxide and temperature. Based on what it would take to kill someone, how much is too much and how much is too little?

In the graph in Figure 1, I have marked the earth's maximum and minimum carbon dioxide levels (while multi-cellular life existed on earth) with two horizontal red lines. As you can see, the level has varied from 5000ppm to 220ppm. Also, the maximum and minimum temperatures are marked with two horizontal green lines showing a high of about 28°C (82°F) and dropping as low as 12°C (54°F).

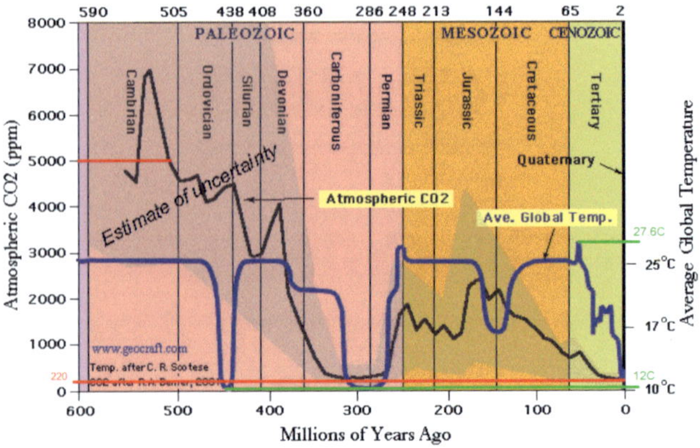

Figure 1 [2]

(Figure 1 has had the detail removed to generalize it in order to give a better sense of the overall view, so don't use it to pick out things about any one time period.)

Based on these measurements, we can say with certainty that life will go on between these extremes: 5000ppm to 220ppm carbon dioxide and 12°C (54°F) to 28°C (82°F) temperature. Our problem here is that the Holocene average is 13.3°C and presently we are at 13.9°C so decreasing temperatures is uncharted territory very close to the practical minimum temperature we have ever endured long term.

But can mankind withstand these carbon dioxide levels? After all, we were not around 500 million years ago when the carbon dioxide was so high.

Here are guidelines set by OSHA on carbon dioxide exposure. OSHA's limits are:

- TWA = 5,000ppm (Time Weighted Average for 40 hrs./week)
- ST = 30,000ppm (Short Term - 15 minutes or less) [3]

The Navy has exposure limits also, mostly for submarines because of concern for proper health and mental stability (they have a lot of nukes on board). Their limits are:

- 1 hour – 30,000ppm
- 24/7/365 - 7,000ppm [4]

Having spent 4 years on submarines I can attest to the fact that 1,000ppm carbon dioxide is not unusual in a submarine.

So we can say with certainty that if the carbon dioxide level stays between 5,000ppm and 0ppm, we humans will do simply fine; no problem here.

And as far as the temperature being between 12°C (54°F) and 28°C (82°F) humans have survived in the arctic and the hottest of deserts. I don't think 54°F to 82°F is going to affect us or most other life forms, and certainly not all life.

Chapter 4

Plants: Our Life Sustainers & Partners

In this chapter I will talk about how plants survive in this environment. And about whether or not plants are important. Sometimes, after we leave school, we forget that life on earth has two distinct but interlocked groups, plants and animals. I think of them as one entity which I refer to as life (excluding the extremophiles).

Life needs energy to survive. Basically, we have two sources of energy, geothermal heat and heat from the sun.

Geothermal heat comes from the earth's core and is produced in two ways. The first is through the decay of radioactive atoms and the second is the constant flexing of the earth's shape due to the gravitational tug of the sun and moon as the earth rotates daily. There is a third heat source that is constantly losing intensity. It is the residual heat within the core that was deposited by impacts as the earth was being "assembled" from smaller parts which is now working its way into space. The heat from these three sources is given off by two methods. The first method is the intermittent expulsion of hot solid material and the production of gases from things such as volcanoes, geothermal vents, geysers, lava flow, etc. The second method is from an extremely small amount of heat released from the surface of the earth since heat always moves towards a colder temperature. This extremely small amount of heat release can't really support much life, except in deep caves. But don't worry, the earth's crust acts as an exceptionally good insulator.

The second heat source is the sun. It sends a lot of energy our way and is the primary source that sets the baseline earth temperature. Plants convert the sun's energy into something we can eat and therefore, live. In return, we give off carbon dioxide which plants use to "breathe" and grow. Not only do plants give us food, they convert our carbon dioxide into oxygen so we can breathe, just as we convert oxygen into carbon dioxide so they can breathe.

This cycle is so mutually tied that if the plants go away so do the animals and humans, and if we go away so do the plants. We understand not having plants to eat will kill us, but we seem to miss what would happen if animals were to go away. Plants don't make carbon dioxide, so if we were to die off the plants would convert all the carbon dioxide into oxygen and have nothing to "breathe." Volcano and ocean out gassing can't produce enough carbon dioxide to make a long term difference. And, conversely, if the plants all go away the animals, even if they had a long term supply of food, would soon convert all the oxygen to carbon dioxide and no longer have anything to breathe. We need each other, desperately.

If you notice in the graph, Figure 1 Chapter 3, the earth's temperature has never been primarily dependent on carbon dioxide. For instance, about 450 million years ago the temperature dropped to 12°C (54°F) but the carbon dioxide stayed around 4,100ppm. Then around 350 million years ago the temperature dropped to 12°C (54°F) and carbon dioxide followed a couple of million years later but never went below 220ppm. Next, about 15 million years ago the carbon dioxide dropped to 220ppm again while the temperature was up to about 20°C (68°F). Then the temperature dropped about 12 million years later. So obviously the carbon dioxide wasn't the major temperature driver; it had to be other natural forces such as the sun, geothermal heat, and other condensable feedbacks such as water vapor, etc.

Chapter 5

Low Carbon Dioxide And Plants

In Chapter 4 I discussed the various environmental factors that have an effect on plants and an overview of the effect of each one on plants. I also discussed how very important plants are to all life on earth including aquatic animals and us humans. What we are really concerned with discussing in this book would be the effect of carbon dioxide that we humans are putting into the air would have on the plants. Is this carbon dioxide good for plants or bad for plants?

There are fundamentally three different types of plants on land, C3, C4, and CAM [5.] C3 plants (most trees, grasses, flowers, bushes, etc.) make up 85% or more of all plants and rely on absorbing carbon dioxide during daylight hours so that photosynthesis can occur as the carbon dioxide is absorbed. C4 plants make up about 5% of all plants and do not require as much direct sunlight because they absorb a large amount of carbon dioxide continuously, store some of it for latcr and through a complex chemical process go through photosynthesis under low light. CAM plants on the other hand make up about 10% of all plants and are plants that survive in hot dry conditions by absorbing carbon dioxide at night through openings which are closed during the heat of the day to prevent moisture loss. After absorbing the carbon dioxide at night, through a chemical process the plants store it and then release it during daylight hours to feed the photosynthesis cycle. There are only two CAM plants used in agriculture, the pineapple and an Agave used to make tequila.

Although we cannot determine all ancient plants characteristics we do know that angiosperms (most present day C3 plants) have been around for 115 million years and some C3 conifers have been around for about 325 million years.

Figure 2 below shows what happens to C3 plants at various ppms of carbon dioxide.

Fig. 2 Representative plants of Abutilon theophrasti (C3) grown at glacial through future [CO2]. All plants were 14 d of age and were grown under similar water, light, and nutrient conditions. These plants were photographed during a study by Dippery et al. (1995). (Photograph is courtesy of Anne Hartley, Florida Gulf Coast University.)

Figure 2 [6]

At about the 100ppm point the C3 plants no longer grow and actually die as there is not enough partial pressure of carbon dioxide to move it into the plants (no tiny lungs to suck it in). At around 20ppm the other two types of plants (C4 and CAM) will lose the ability to absorb carbon dioxide and they will die as well.

So, when the carbon dioxide gets too low, the plants don't grow as big. Which puts lack of food pressure on the animals which causes them to die. The balance point of our present conditions is when the decreasing number of animals leads to increased oxygen content in the air. This allows more animals to live (and produce more carbon dioxide) which helps the plants, but low temperatures prevent animal populations from growing such that they can produce more carbon dioxide then the plants use and you wind up in a steady state of low temperature and low carbon dioxide balance. In Chapter 8, *How Is Oxygen Doing* it is stated that for the last 40 million years the balance point has resulted in oxygen steadily decreasing ever so slightly (see Figure 5 in Chapter 8).

With carbon dioxide and temperature this low, anything lowering the temperature or carbon dioxide could literally end all multi-cellular life on earth (except most of the extremophiles).

So, let's not go lower than 350ppm carbon dioxide!

Chapter 6

Decreasing Temperature

In the introduction I stated cold kills, but what says cold really kills more than heat?

It seems cold actually does kill more humans. Below is a graph of the all human deaths per day per month from the years 2001 – 2007 in the United States.

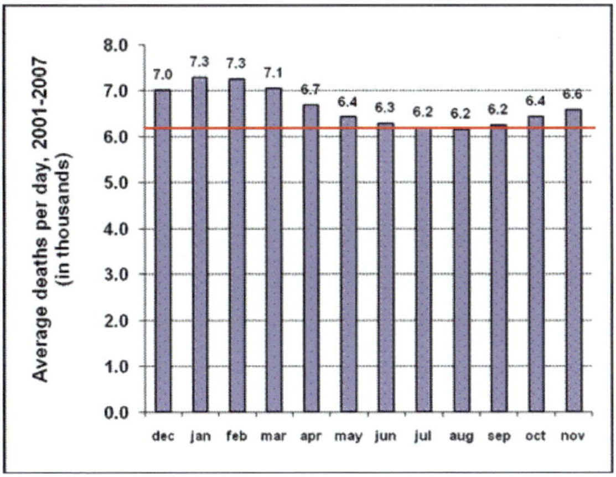

Figure 3 [7]

You will notice that in the warm months of June, July, August, and September there are about 6,300 deaths per day in the US. In the colder months, October through May, there are about 6,800 deaths per day. In total, there are about 156,000 excess deaths during the cold months in the US per year. I cannot find data for the rest of the globe, but based on population numbers alone, there COULD be as many as 2.3 million deaths per year worldwide due to cold. A significant number!

Considering that in the Medieval Warm period the earth's population went from gaining 13 million people per 100 years for 300 years to gaining 30 to 59 million people per 100 years for 400 years it is obvious that the past shows warming is good for humans[8].

Morally one could say that people have been dying from cold for a long time and there is nothing unusual about it. That is the way of nature. And in one sense that is true, but as human beings we have an obligation to try and prevent unnecessary deaths.

But does dropping temperature affect life other than humans right now? Well, if you live in an area like North Dakota you realize that more animals die in the winter when the weather gets colder. When the snow melts you can see the dead carcasses in the fields. This happens because animals are not evolved to handle cold weather as well as warm weather, and when unusual conditions occur there are many more that die. This is also why animals have their babies in the spring rather than the fall, so the babies can get stronger before the cold weather hits. Animals that hibernate generally hibernate in the winter, not the summer. All of this is brought about because of the laws of physics; we need energy to live and energy loss in colder weather reduces energy for living.

The degree of cold an animal can withstand strictly depends on how the animal has evolved. Present day animals that live strictly in the desert, even though they tend to be nocturnal and active during the cool hours of the night, still do not survive in Wyoming because of the relative cold. Granted, some of these animal deaths are not due to exposure to the cold, some are due to more contacts with humans as they look for food and shelter or such, but the fact is - all of that comes with colder temperature and cannot be ignored until we can bring the dead back to life.

It is rather obvious that animals, including us, are more susceptible to dying when the temperature is below what we have evolved to handle due to the laws of physics. Therefore, it is also obvious that dropping temperatures are going to stress the worlds animals (including us) much more than rising temperatures. And if the temperature drops, the rate of evolving must be high to keep up. Energy loss is an immediate problem, not an evolutionary time period problem.

The real killers are mass extinctions. So, in the past was it cold, warmth, or neither that contributed to mass extinctions? Figure 4 below, with considerably less smoothed plots of temperature and carbon dioxide, shows extinction events and the things that could contribute to those events[9, 10, 11, 12, 13, 14, 15].

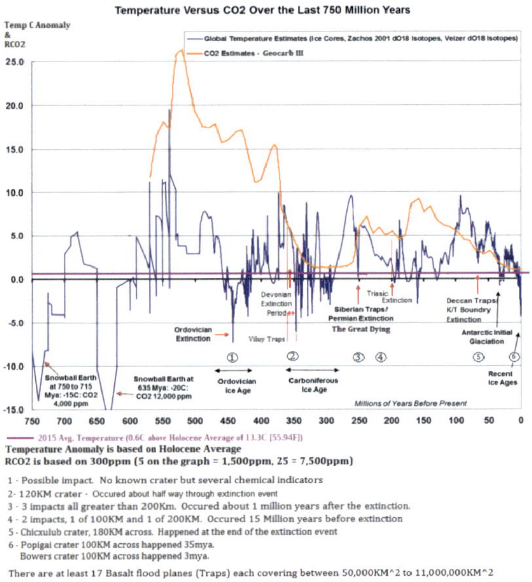

Figure 4

The first thing that stands out in Figure 4 is that there is no correlation of absolute temperature to event. So, just because it is generally warm or generally cold does not seem to be important.

But all the extinctions occurred when the temperature was dropping, never when it was rising. This fact is important, but you also must consider that there are many times when the temperature fell and no extinction occurred. So, there must be something else besides just temperature dropping that contributes to the extinction events.

Carbon dioxide does not appear to be a major factor in extinctions either, it can be going up or down and has varied between 5,100ppm and 300ppm during extinctions. So, there is no direct correlation to carbon dioxide ppm and no direct correlation to changing levels, just as there is no direct correlation to absolute temperature.

There were two periods in Earth's history known as "Snowball Earth" periods. During these events the earth would have looked like a giant snowball because there was ice covering the vast majority of the land and the ocean and the average temperature of the earth was 3°C (5.4°F), 10°C (18°F) lower than the average temperature of the Holocene Epoch (The last 11,500 years). During this time there was no animal life on earth, possibly extremophile and single celled life forms, but no "normal" multi-cellular life forms. Multi-cellular life forms began in earnest at about 540 million years ago during the Cambrian explosion when the earth warmed up to 23°C (73.4°F), 10°C (18°F) above the Holocene average and the carbon dioxide was approximately 4,500ppm.

This can't be the entire story though.

Chapter 7

Combinations?

So, what are some of the contributing factors that may over stress plants and animals to kill them off? A look at Figure 4, with major "earth shattering" events added in, may tell us. Let's look at the five mass extinctions.

The first extinction was the Ordovician period about 450 million years ago in which about 52% of all species became extinct. The temperature was dropping, carbon dioxide was rising, and there was an impactor that hit the earth during the extinction. No crater has been found yet, but the same chemical signature was present when all the other impactors hit.

The second extinction was the Devonian period which started about 374 million years ago and ended about 354 million years ago. During this time 70% of invertebrates became extinct (almost no land, plants, or animals were affected). The temperature and carbon dioxide were dropping during this time. Once again, an impactor hit and created a 120 km crater about 364 million years ago. Unlike all other extinctions, this one took about 20 million years, so it was extremely slow and was mostly just species dying off and others replacing them. This appears to be more of an "evolutionary extinction" where the species dying out were just not able to evolve to the lowering temperatures and other environmental factors.

Next, we have the big one. The Permian/Triassic period extinction happened when the temperature was dropping, carbon dioxide was dropping, the Siberian Traps had been spewing lava for quite a while, and, finally, the oxygen in the air and water was dropping drastically. There were four impactors that all produced 180 km or larger craters. These happened somewhere between one million years before the extinction occurred to during the extinction. The accuracy of the dating prevents saying for sure if they happened before or during.

Oxygen levels dropping had a major effect on this extinction. The atmosphere dropped from about 30% oxygen to about 16% oxygen by the time the extinction was complete. Animals do not do well without oxygen, and 16% oxygen is equivalent to an altitude of 9,200 ft. There were multiple problems: Dropping temperature, the Siberian Traps putting carbon dioxide and contaminants into the air, and the oxygen dropping both in the atmosphere and the ocean. This wiped out 90% of all aquatic species, 75% of all land animals and plants, and 83% of all insects.

This event, which began at a temperature of about 5.4°C (9.8°F) above what it is today and carbon dioxide at 1,800ppm, almost killed off all life on earth. The oxygen makers, plants, were dying off thus reducing the oxygen which reduced the number of animals which then reduced the carbon dioxide and so on. This happened as the Siberian Traps continued flooding the earth with lava and the earth continued to cool until it was about 6.6°C (12°F) colder than today. It took almost ten million years for life to recover from this event.

The fourth extinction was the End Triassic Extinction period which occurred about 201 million years ago. It wiped out 50% of all species on earth. Temperature was going down and carbon dioxide was going up. The only thing happening during this time was the splitting apart of the continent, known as the Pangea and Central Atlantic Magmatic event (1.5 times larger than the Siberian Traps but under water). The splitting of the continent and major volcanic activity only took about 40,000 years. There are multiple theories as to the cause of extinction this time, varying from carbon dioxide and methane warming up the atmosphere to ocean chemistry changes. It wasn't a mass extinction, just creatures filling in niches created by the splitting of Pangea and being better adapted to take over.

The fifth and final major extinction was the one that killed the dinosaurs and is known as the K-T Boundary extinction period. It happened 65 million years ago when temperature and carbon dioxide were dropping, the Deccan Traps were spewing out lots of lava, and there was an impactor that hit in the Chicxulub area of Mexico that created a 180 km wide crater and killed off about 80% of all species.

The only two consistent characteristics that have shown up every time during an extinction are geological activity and cooling temperatures. Geological activity, as of now, cannot be stopped or controlled once it starts. Decreasing temperatures may be something we are able to stop.

Of our two culprits it was never the absolute temperature of the earth or the absolute ppm of carbon dioxide or how the carbon dioxide is changing that factored in the extinctions. It was only the characteristic of decreasing temperature.

We know that it is not just a cooling climate that causes extinction because the climate has cooled many times with no extinction taking place.

We know that it is not just geological activity that causes extinctions because there have been 17 known major Basalt Flood events[17] in earth's history and only five major extinctions. And some of those flood events were larger than those during extinctions.

We know that it is not just the outflow of lava and its associated gasses that creates an extinction. Lava flowing out onto the surface will kill a lot of "local" species that do not manage to escape the boiled lakes, wild fires, or lava, but as far as the entire planet there is more than enough time for the mobile animals to get out of the way.

And extinctions occur whether carbon dioxide is rising or falling, whether it is high or low.

A major threat is created only when there are other factors that add to the cooling climates. Factors such as geological activity, impactors, loss of oxygen, or a combination of these things. And these combinations are the ones that kill off the mobile animals such as dinosaurs, fish, etc.

Addressing impactors is now becoming possible. NASA is continuing work on the NEO (Near Earth Objects) project to identify asteroids and comets whose paths cross with or come close to the earth. So, as we progress, we should eventually have a fairly good system to give an early enough warning that even with our primitive technology we could redirect many impactors.

Next, we need to address the loss of oxygen in the water and atmosphere. That's going to be a tough one. Nobody has addressed this issue, so there is no information as to what can be done about it. An issue for serious consideration in the technology and scientific community.

In conclusion, we have found that true life-threatening things combined with cooling climates are the killers:

1. Loss of carbon dioxide such that plants are no longer productive.
2. Loss of oxygen in the atmosphere and oceans such that animals are under extreme stress.
3. Geological activity.

The colder our planet, the more animals and humans die due to the cold climate.

So, let's not deliberately decrease the carbon dioxide or cool the planet off or lower oxygen to a dangerous level!

Chapter 8

How Is Oxygen Doing?

In Chapter 7, Combinations, I discussed how we do not want to lower oxygen to a level which stresses animals. But the question is, is it presently going up or down? Below is Figure 5, a graph of the oxygen percentage of air over the lifetime of animals. Variable sedimentary rate values have been removed from the chart.

As you can see, oxygen has varied widely over the lifetime of animals. Some geologists have found that the oxygen level was at or below the 15% mark during three of the major extinctions. It was at about the same level as today during one event and even higher during another. So, picking 15% based on this seems to be more arbitrary than rational since these low values are only dominant because the oxygen has been low for about ½ the time animals have been on the earth. The probability that there would be three out of five events during the low oxygen period in the life of the earth would not seem to be statistically significant.

The vertical lines in Figure 5 represent the same years as the major extinctions shown in Figure 4. The horizontal solid line is the 15% oxygen line and the dashed line above is the 20.94% line equating to where today's oxygen is.

NIOSHA limits workspace entry oxygen to no less than 19.5% to allow sufficient oxygen such that self-extraction is not impaired by oxygen starvation[18]. In other words, having to climb out of an enclosed area wearing safety equipment won't be impaired because of low oxygen. The "death zone" when mountain climbing (a purely arbitrary distinction based on general knowledge not research) is about 6% oxygen. We can be fairly sure that if we limit ourselves to no less than 19.5% oxygen animals of the present will survive with no harmful effects.

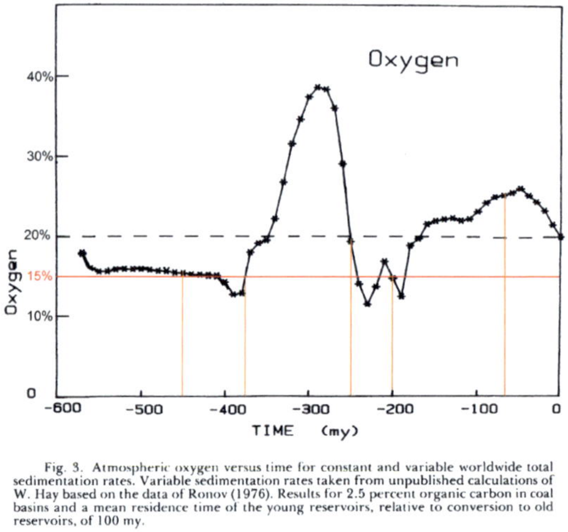

Fig. 3. Atmospheric oxygen versus time for constant and variable worldwide total sedimentation rates. Variable sedimentation rates taken from unpublished calculations of W. Hay based on the data of Ronov (1976). Results for 2.5 percent organic carbon in coal basins and a mean residence time of the young reservoirs, relative to conversion to old reservoirs, of 100 my.

Figure 5 [19]

Oxygen right now, as well as for the last 40 to 50 million years, has been dropping. In the past it was dropping at 0.0005ppm/yr. and presently it is dropping at a rate of about 4ppm per year (0.0004%). In 2008 the oxygen was at 20.94%, so it would have to drop 1.45% to say it has dropped enough that we need to do something about it. At 0.0004% per year that will take 3,625 years.

Therefore, we have at least until the year 5,633 CE before it becomes a problem. We do not have enough fossil fuels to burn for that long[20], so it is not a significant concern. We should watch it to see if increasing plant production slows the rate of oxygen depletion.

We *can* say that oxygen and its rate of depletion is not a problem.

Chapter 9

How Are The Plants Doing?

Are there any other benefits to higher carbon dioxide, such as larger crop yields?

First, what would we expect plants to do? Research by Taiz & Zeiger, as pictured in Figure 6 below, shows that C4 plants (high plains type plants, tumbleweeds, some grasses, corn, sorghum, etc.) are all but maxed out at present in their ability to produce plant material. They have only risen slightly (~1%) since the preindustrial 289ppm carbon dioxide level, as shown by their net photosynthesis line being flat when carbon dioxide is more than about 350ppm.

C3 plants, on the other hand, have just gone over the 50% mark in their production of plant material since the industrial revolution.

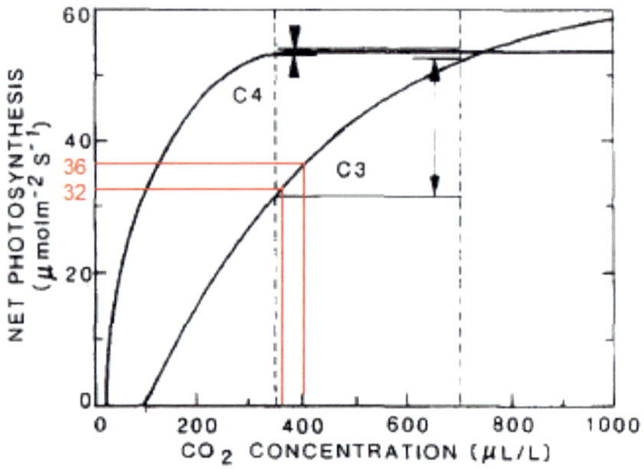

Fig. 3. Classical net photosynthetic curves for C_3 and C_4 species (Taiz & Zeiger, 1991). Dashed vertical lines at 350 and 700 μl/litre mark the current CO_2 level and the doubled concentration predicted to be reached sometime late in the next century (Houghton *et al.*, 1990). Arrows indicate incremental rise in net photosynthesis due to the CO_2 doubling (Kimball *et al.*, 1993).

Figure 6 [21]

The second factor is that plants also respond to temperature. The average temperature change in the land area of the earth from 2002 to 2013 was only 0.1°C (0.18°F)[22]. Based on this, the effect of temperature over the 2002 to 2013 range would be effectively zero, although an analysis on a yearly basis may show some gains and/or losses.

And the third factor is environmental changes, such as humidity, cloud cover, the methods mankind uses to grow food (such as better weed control, better soil nutrient control, better pest control), etc. These factors are hard to quantify due to lack of reliable data about actual implementation and actual environmental factors.

Based on Figure 6 we do know that C4 plants will not change production very much with carbon dioxide changes in the range we are working with. Thus, we should be able to subtract 1% from their total gain in yield to get a good estimate of what percentage the other factors have changed the yield.

We have records of worldwide crop yields from 2002 to 2013[23].The average carbon dioxide in 2002 was 373ppm and in 2013 396ppm for a net change of 23ppm or 6.2% [24]. Graphing the carbon dioxide levels for those years onto Figure 6 we find that the net photosynthesis should have changed from 32.7 to 36.6, or 10.4%. This would indicate we should expect worldwide yields of C3 plants to grow about 2/3 as fast (in percentage) as the carbon dioxide rises. The actual yield is graphed in Figure 7.

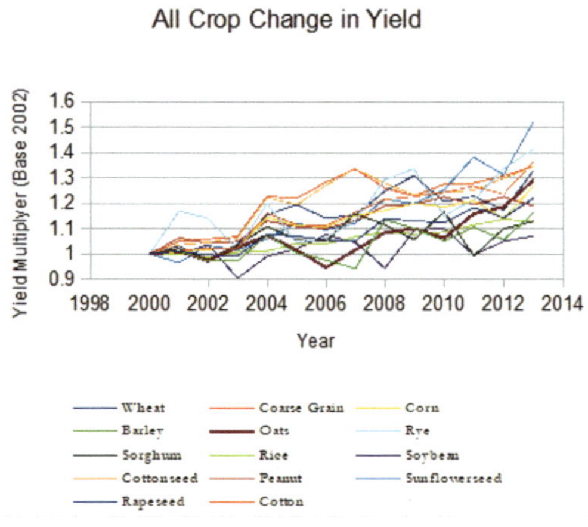

Figure 7 [25]

Our records only have two C4 plants in them: corn and sorghum. Their average rise was 20%. Subtracting out the 1% for carbon dioxide rise shows 19% of their gain was due to other changes not associated with carbon dioxide.

C3 crop plants yielded 28.2% more on average. Assuming that the 19% of C4 change that was not due to carbon dioxide changes also applies to C3 plants we see they actually rose 9.2% due to carbon dioxide alone. Considering that our difference from the theoretical 10.4% to the practical 9.2% is off by only 1.2%, it would be reasonable to assume the total effect of future rise in carbon dioxide at a minimum will continue along the same percentages as shown in the graph by Taiz & Zeiger. The actual increase in yield per hectare will probably begin to slow down as we start to run out of the efficiency gained by external changes. In other words, we will get to the point where things like the improvement in fertilizers overruns the ability of the plant to grow. At that time we will no longer have more gain in yield year to year because of fertilizer improvements, etc.

It is obvious that even though the entire earth is warming, and carbon dioxide is rising, the sum total effect is that the plants are growing more food per hectare as we are growing more people.

So now we know that as the carbon dioxide rises the temperature will rise slightly and crop production will also rise at least as fast as Figure 6 shows and probably faster.

Figure 7 is a bit of a mess because there are so many different crops and not all crops perform the same as the general graph in Figure 6. The four largest crops by acreage are wheat, corn, rice, and soybeans. Their yields are plotted in Figure 8.

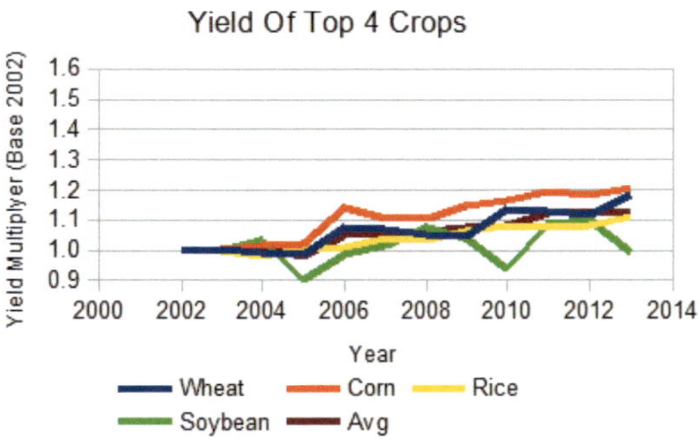

Figure 8 [25]

Soybeans gained the least in yield over the 12 years of data. The average gain in yield for these four years was 12.6%, less than ½ the C3 average of 28.3% but still more than the carbon dioxide induced change of 10.4%. This may be true because these crops have received the most attention over the last several thousand years because they are the crops mankind has lived from. The soil conditions, when to plant, when to harvest, the machinery used to harvest, and the pests are all well understood by the agricultural groups throughout the world. Thus, their gains would be expected to be the least because they were already pretty much close to the limits on everything and are primarily affected by the carbon dioxide rising. Their lower gain in yield over these 12 years will become the norm for the gain in other crop's yields as time goes by.

They are also an indicator of what plants can support as far as numbers of people on the earth. If plant yield only goes up 1% per year, our population should go up a maximum of 1% per year. Presently our population is rising at around 1.13% per year[26]. The four major crop's yield has been increasing by about 1.24% per year. This ratio has not increased our worldwide reserves of food and we are running at the break even point between waste/spoilage and eating the food. If mankind had not been putting carbon dioxide into the air, then the crop yield increases would have been down from 1.24% per year to 1.20% per year. This means we would have to find 0.04% of the earths present agricultural acreage of 4.62 billion acres[27] which is 1.848 million acres (2,888 square miles) more every year to plant crops just to break even[28] and then hope nothing major goes wrong for any period longer than our reserves can hold out.

In addition, we as humans will have to take larger and larger chunks of land away from natural habitats occupied by other animals in order to feed ourselves. Or we as humans will have to limit our birth rates. Both of these limits seem like impossible tasks compared to possibly losing some non-productive beach front property.

Chapter 10

Evidence of Warming Extinctions?

We must ask ourselves if rising temperatures have ever caused a problem for life on earth? There is no absolute evidence that the conditions at the time of the five major extinctions did not include temporary temperature rises of some sort for some intermittent duration. The temperature record does not indicate that rises were a major contributor to the stresses the animals were under when the extinctions occurred.

This also makes sense because, as we know, for a large part of the animal and plant species to have died off would have required that the majority were stressed. The majority of the world's animals are warm adapted rather than cold adapted. It is also easier for cold adapted animals to survive in a warm environment simply by shedding their warming "coat." And by coat I am referring to adaptations such as hair or fat or other diverse things animals have developed to adapt to cold weather.

Warm adapted animals must change their genome to produce a warming coat and this evolutionary change would take a very long time.

It is also important to note that none of the major extinctions occurred while the average temperature of the earth was rising. In other words, it is apparently easier to shed an existing coat than to develop a new one based on your present genome. This would make sense since deactivating an existing gene would be more probable in the short term as opposed to changing the gene to produce a coat.

Realistically, warming up, like any change, will result in some loss of species. But those species would have to be barely adapted to the existing conditions to begin with since our rapid seasonal changes are so extreme. If the animal cannot withstand a few degrees permanent change it would not in the long run survive the cyclic changes in seasons themselves, like the 100 year floods, the droughts, winters, summers, and such.

Although there is no absolute data saying temperature rises do not contribute to mass extinctions, the data I have says it does not.

Chapter 11

What About The Rate Of Climate Change?

What about the rate of climate change? Can it cause extinction? Extinctions like the Devonian period extinction took millions of years, and the K-T boundary extinction period's final "nail in the coffin" was probably as short as several days to several decades[29]. The answer is that the rate of climate change is a factor depending on how massively the environment is disrupted. In other words, how many different environmental factors are changed and how rapidly?

In the case of the K-T boundary extinction, global temperature was already dropping. The Deccan Traps were changing the chemistry of the air and water with carbon dioxide and other chemicals as well as changing the land in a large area. The area covered about 1.5% of the earth's land surface (2,000,000 km^2). And at some point, all the environment changed rapidly and those not already adapted to the new conditions did not survive.

Therefore, we can say that the rate of change is important only in that it depends on how many factors are changing at once and how the animals adapt to the final result.

Considering that the adaptability of animals is an especially important factor, it would be good to know what the natural rate of change the present animals on earth have experienced.

From the Greenland temperatures derived by Alley[30] we can say that Greenland has experienced a maximum rate of temperature rise of 0.356°C (0.64°F) per decade. The peak rate happened about 3,300 years ago, long before man was doing much. The work of Box[31] showed that Greenland has a 1.6 times greater magnitude of variability than the northern hemisphere. And according to data retrieved from the Hadcrut 4 database[32], the northern hemisphere is 1.289 times more responsive than the earth as a whole. That means Greenland responds 2.06 times faster than the globe as a whole . That would indicate that for the entire globe, over the last 10,000 years when there was no measurable man made carbon dioxide, the earth as a whole had a maximum global rate of change of 0.172°C (0.31°F) per decade.

Looking at the Hadcrut 4's data we can see that during the period 1975 to 2005 the global temperature rise was a steady 0.19°C (0.342°F) per decade, or 10% higher than has been measured in the last 10,000 years. Taking into account that modern measurements are not smoothed, but the 3,300-year-old Greenland data is inherently smoothed, 10% difference is easily within the possible values of a not smoothed Greenland reading.

If it is actually the carbon dioxide that is primarily warming the globe, considering that carbon dioxide heating is a logarithmic function, the rate of warm up should begin to be noticeably less as time goes by.

It is safe to say that the rise we are experiencing is not beyond what is natural within the time frame that man has been on the earth. Therefore, it will make no significant difference to the plants or animals of this planet.

Chapter 12

Sea Level Change

Sea level rise is the last of the big concerns. Throughout earth's history the ocean has risen as much as 400m [1300ft][33] (max – probably closer to 300 m [984ft]) several hundred million years ago.

There are several problems with trying to say what the sea level was in the very distant past. In the very, very distant past the amount of water present on the surface of the earth varied. In addition to this, the contour of the ocean floors and continents, especially the ocean floors, involved a lot of guesswork. The continents themselves could have protruded from the theoretical true shape of the globe more than they do now, or about the same, or even less. And the ocean floors could have had larger rifts in them or fewer rifts in them. In Figure 9 you will notice that the marks for sea level are indicated by high and low instead of meters. And most importantly, no geologist is willing to give you an exact value for what the sea level was.

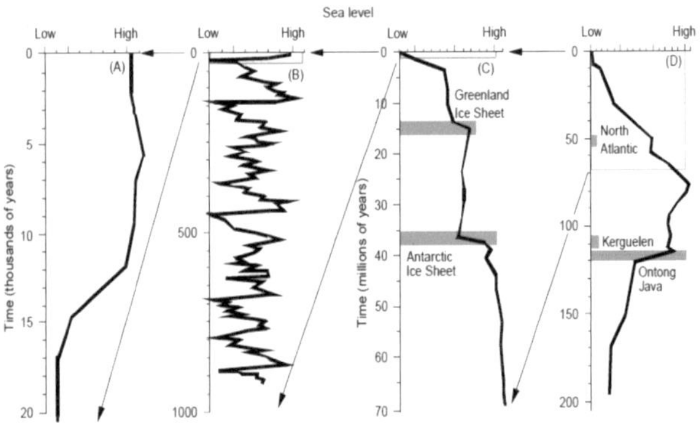

Figure 1 (A) Estimates of sea level change over the last 20 000 years. Amplitude is about 120 m. (B) Northern Hemisphere glaciers over the last million years or so generated major sea level fluctuations, with amplitudes as high as 125 m. (C) The long-term oxygen isotope record reveals rapid growth of the Antarctic and Greenland ice sheets (indicated by gray bars) as Earth's climate cooled. (D) Long-term sea level change as indicated from variations in deep-ocean volume. Dominant effects include spreading rates and lengths of mid-ocean ridges, emplacement of large igneous provinces (the largest, marine LIPs are indicated by gray bars), breakup of supercontinents, and subduction of the Indian continent. The Berggren *et al.* (1995) chronostratigraphic timescale was used in (C) and (D).

Figure 9 [33]

There are some very subtle things that can change the sea level over time, such as the Indian continent subducting under the Eurasian continent. As India moves north, its plate is pushed under the Eurasian continent which raises the Himalayas and leaves a void behind which the oceans fill resulting in sea level drops. You cannot see this happening as it takes millions of years to show any real shoreline change. And closer to home, along the San Andreas fault, the Pacific plate is sliding in a north-south manner past the North American plate yet is making no difference in sea level.

Has anyone given a serious estimate of the sea level rise vs. land shifting? No one that I can find, because apparently no one is sure.

Based on the IPCC model estimates, between now and 2100 the oceans should rise somewhere between 15 cm (6 inches) and 95 cm (37inches)[34].

And when considering this we must realize that the IPCC model estimates that over the last 100 years the sea level has changed by a low value of -19 cm (-7.5 inches) to a high of 37 cm (14.5inches). Yes, a drop in sea level. And based on tidal gauges over the same 100 years as the model, the low estimate of sea level rise is 10 cm (3.9 inches), and the high estimate is 20 cm (7.8 inches). The thing to consider is the tidal gauges have a 10 cm (3.9 inches) difference low to high and the IPCC model has a 56 cm (22 inches) difference for the same time period, indicating that the IPCC model over estimated reality by a factor of 5.

Chapter 13

Five Hundred ppm

Figure 10 shows the results of Dana Royer's research into the latest glaciation period[35]. This graph is about the best illustration possible.

Each circle and triangle represent a carbon dioxide reading based on the indicated method used to get the value. The width of each circle and triangle represents hundreds of thousands of years. They are NOT daily, monthly, or yearly readings!

Notice that because the carbon dioxide has been incredibly low the last 33 million years there was ice age after ice age. Before that, when the carbon dioxide varied up and down quite a bit, there were "signs" of ice, but not necessarily ice ages.

Royer points to the fact that above 1,000ppm carbon dioxide you get no glaciers. Between 1,000ppm and 500ppm you get intermittent short-lived glaciers and below 500ppm you are in an ice age with major glaciers coming and going with the glacier free times much less than the glacier times.

Although it is easy to draw the conclusion that carbon dioxide or its effects were the cause of the loss of glaciation, we all know that correlation is not causation. Is there anything that would indicate it was not primarily the carbon dioxide that caused the change?

A brief glance at Figure 4 in Chapter 6 shows that carbon dioxide in the 66 million years ago to 33 million years ago range was higher than our last 33million years ago when we suffered ice ages. But the temperature was also higher, so was it just that something raised the temperature which also

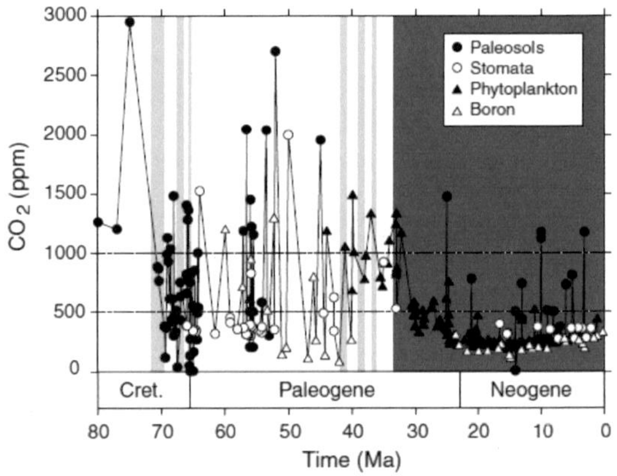

Fig. 4. CO_2 and temperature records for the late Cretaceous to present day (80–0 Ma). The CO_2 record is derived from Fig. 1B. Cold periods with strong evidence for geographically widespread ice are marked with dark shaded bands. Cool-to-cold periods with indirect or equivocal evidence for ice (see Section 2 for details) are marked with light shaded bands; such periods supported by only weak evidence are annotated with a question mark. The horizontal dashed lines at 1000 and 500 ppm CO_2 represent the proposed CO_2 thresholds for, respectively, the initiation of globally cool events and full glacials.

Figure 10 [35]

raised the carbon dioxide? To find the most likely answer let's look at the contribution of carbon dioxide to temperature in the two time periods of 66 million years ago to 33 million years ago and 33 million years ago to 0 million years ago.

The present best guess estimates of carbon dioxide doubling temperature is between 3.25°C (5°F) and 1.5°C (2.7°F). Using the average of 2.375°C (4.275°F) we can account for 11% of the heat over the previous 33 million years and 44% over the range of 33 million years ago to 66 million years ago. Therefore, it is logical that the carbon dioxide is not the major driver of the temperature even over extremely long million-year ranges. It only provides about 32% of the heat. The other 68% must come from variations in other sources.

Between 52 million years ago to 60 million years ago carbon dioxide was below the 500ppm point for the vast majority of the time and yet no glaciation took place. The difference here is that there were a bunch of spikes of carbon dioxide during this time period which we have not experienced for the last 33 million years. Whether these spikes were caused by geological activity or plant/animal imbalance is not presently known. The fact is they occurred and apparently either they, or more likely them in combination with another heat source, made the environment more stable and livable for the animals that were there. These spikes in carbon dioxide level during the 65 million years ago to 33 million years ago with no significant extinctions indicates that the animals that survived and evolved after the K-T Boundary extinction and up to the present day have not been severely impacted by these rises in carbon dioxide level.

Chapter 14

Orbital unknown and 560ppm

Figure 11 is a graph of the insolation (the amount of energy from the sun arriving at a particular spot). It is represented by i_0 on the graph at 65 N in June or July for the previous 500,000 and next 500,000 years. This is the most reliable present indicator of a glaciation start we have[36]. The top of the graph is labeled pCO_2, µatm. We know this as carbon dioxide in ppm. The upper left half of the graph is what was measured as the past carbon dioxide in the atmosphere. The right half has a green, a blue, a purple, and a red line showing how carbon dioxide varies over time if various tons of carbon are burned.

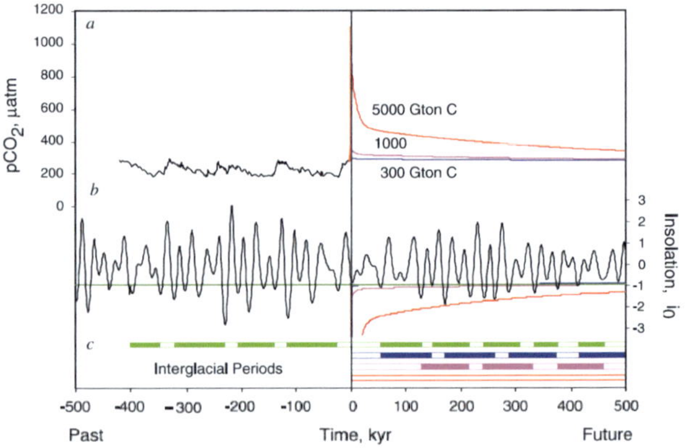

Figure 11 [36]

Below the carbon dioxide lines there is a squiggly black line showing the insolation we have received and will receive in the future at 65° North. A green line runs through it that is the insolation point where huge glaciers would have been generated if we had never added carbon dioxide to the atmosphere (280ppm).

There are also blue, purple, and red lines which are inverted and mapped onto the squiggly line based on their relationship to the green line. When these lines go above the black insolation line you trigger the start of glaciation. How long the glaciation will last depends on many factors, not just the insolation line going above the trigger point.

In the following discussion all the white squares represent when there are no significant glaciers (except Antarctica and Greenland). The colored squares represent when the earth has gone into or will go into glaciation. During these times the earth will have 1¾ mile thick glaciers covering most of the northern hemisphere.

> Note: 1 Gton C is 1 GigaTon (billion tons) of carbon burned. Each Gton of carbon burned produces 3.67 GTon of carbon dioxide (oxygen adds weight). And since the atmosphere weighs 5.3 million GTon, each Gton of carbon burned produces 0.69ppm of carbon dioxide.

Below these insolation lines are four rows of colored squares. The top row shows green squares for glaciation periods. The left half is the past and the right half is what will theoretically happen if humans never add carbon dioxide to the atmosphere.

If we add 200ppm (the blue 300Gton line) of carbon dioxide to the atmosphere while ignoring natural changes in carbon dioxide, then the start of glaciation line (blue) which is almost completely hidden by the green line shows the 560ppm carbon dioxide curve. It goes up to 560ppm and then in a few thousand years drops back down to the 280ppm line. You can't see the rise and fall because the scale on the bottom covers 100,000 years for each tic, and several thousands of years is a really thin line! A small, almost invisible little bump at the black vertical line crossing the green line indicates the change in insolation to start glaciation after adding 200ppm carbon dioxide. This guarantees we will not go into glaciation right now. And continuing to the right we get to the blue squares of theoretical glaciation. If all this is correct, then the next glacier won't begin for 50,000 years.

Instead of 300 GTon, if we inject 1000 GTon (690ppm) of carbon into the atmosphere the result is the purple lines. Carbon dioxide goes to 1090ppm and we see the purple line running through the insolation lines. To the right are the purple squares of theoretical glaciation. If all this is correct, then the next glaciers won't begin for 125,000 years.

Beyond that, if we inject 5000 GTon of carbon, the carbon dioxide goes to 3,810ppm and we get the red line of insolation which prevents glaciation for at least 500,000 years.

It will take a full 400,000 years until the orbital changes of the earth will again be stable enough to prevent glaciers.

The right-hand side of the graph, the calculated insolation values, seem to follow the actual glaciations in the past. The point here is that these are theoretical values and as such could be off just a little bit. Archer & Ganapolski state in their paper, "Paillard's i0 was a somewhat arbitrary choice within a wide range of acceptable values, however, and a slight change in i0 could easily tip the simulation into the onset of glaciation now rather than in 50kyr. It appears that the natural evolution of the next few thousand years is a close call whether to glaciate or not, an issue of subtle differences in models rather than a fundamental difference between them."[36] In other words, we are on the razors edge of going into glaciation.

If you look at a closeup of the center of the insolation as it is today, you will see in Figure 12 that we have approximately another 1,000 years of decreasing insolation due to orbital changes. But we have no way to determine the effect of the sun's variability on this value. The total energy from the sun can go up or down, putting us closer or farther away from the tripping point. At this time, it appears to be giving off relatively less and less energy and may go into a lower energy state for a considerable period of time.

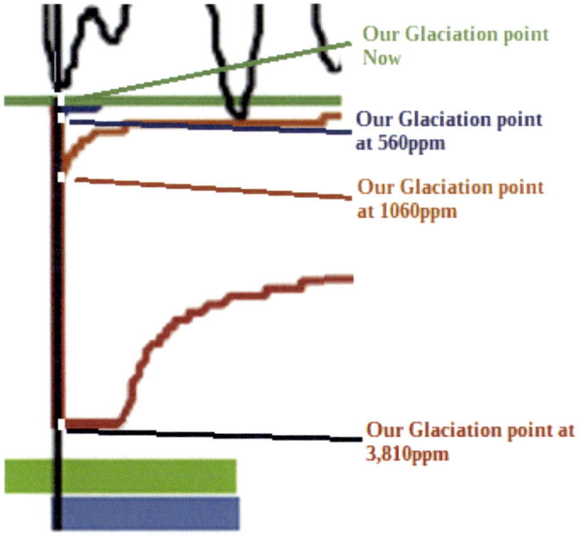

Figure 12

The model below done by Archer & Ganapolski predicts that "…at 560 µatm the model will not glaciate at all within a reasonable range of orbital eccentricity."[36] This model plus the unrelated real-world work of Dana Royers discussed in Chapter 13 point to very strong evidence that above 560ppm large glaciers will not form. Thus, the safe conclusion according to both real world and model data would point to a 560ppm value (or slightly higher to ensure that we do not enter into an ice age).

Chapter 15

Present Day Geological Activity

Presently there is geological activity helping us prevent glaciers. Looking at Figure 13, a graph of the seismic activity in the regions of the continental plates that are spreading apart, we see that after 1994 there was a drastic increase in the amount of seismic activity. The activity reached its peak in 2014 [37].

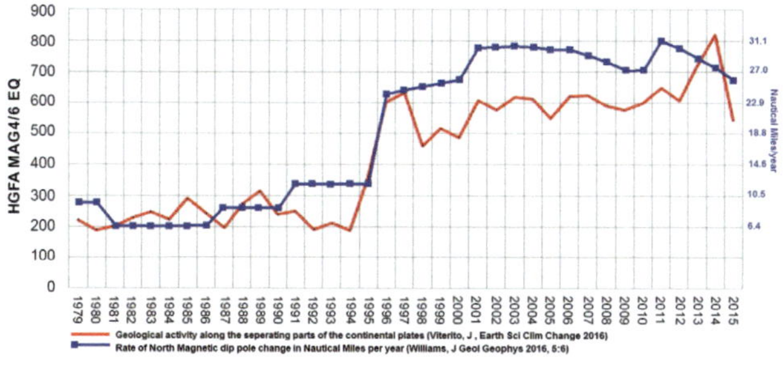

Figure 13 [37, 38]

The seismic activity represents tremors in the magnitude 4-6 range and only those occurring where the ocean floor is pulling apart. Not small, but not huge like undersea volcanoes. More like undersea cracks opening and allowing lava or gasses to flow out rather than erupt.

Shown along with this is the speed the north magnetic pole was moving[38]. It too made a drastic change around the same time. A cursory look at the graph seems to show that each time the seismic activity increases the speed of the magnetic pole increases and as the amount of heat released through seismic activity decreases the speed also slows down. This speed change would be produced by the change in the distribution of magma at different temperatures below the crust as it escapes into the oceans.

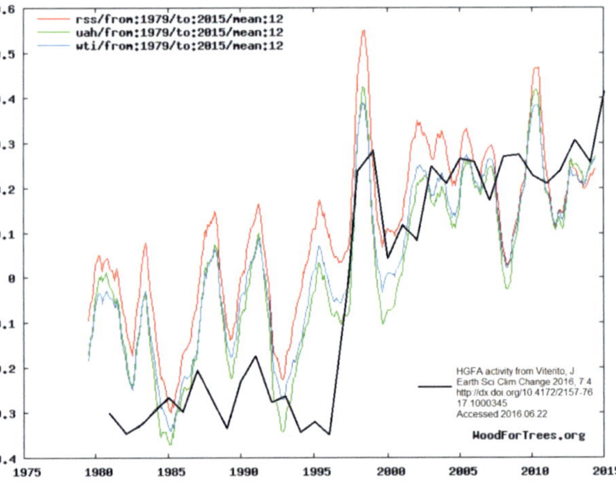

Wood For Trees Temperature Index, RSS, and UAH uncorrected temperature readings with HGFA seismic activity plotted advanced 2 years. The beginning reading at the Temperature year 1981 is the HGFA activity in 1979. The Temperature 2015 reading is the HGFA 2013 reading.

Figure 14 [38, 39]

Figure 14 shows the earth's temperature as measured by satellite (RSS & UAH) with the HGFA seismic activity delayed two years and then plotted with the temperature of the earth. The two-year delay is based on the fact that deep ocean warming takes a while to reflect to the surface. The correlation is strikingly good, much more so than the carbon dioxide correlation.

To check this correlation, Figure 15 has the extent of ice on the Great Lakes shown with an inverted seismic activity graph also delayed two years.

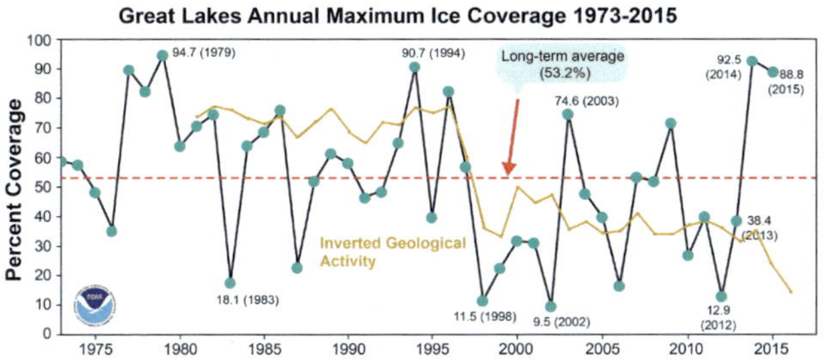

Figure 15 [37, 40]

If this seismically released heat is the cause of the ice changing then an inverted graph should look just like the ice graph since low heat output is now a high value on the graph. Low heat should produce a higher amount of ice and vice versa. It does not depend on the particular temperature reading; it is just a measurement of the amount of ice. From the left hand side to 1997/1998, the amount of ice was generally above the average indicating it was cooler. From 1997/1998 to 2015 the ice coverage was lower indicating a higher average temperature. This follows the seismically released heat. You will notice that on both sides of the 1997/1998 line the ice sometimes is off from its general trend, either high or low. The ice on the Great Lakes is weather generated ice which can be drastically different due to where the jet stream is. Just how much effect the jet stream has on the ice is influenced by the earth's average temperature, as well as the position of the magnetic north pole and various other short-term environmental factors such as dust, humidity, etc.

So, we know there are two indications that the earth's average temperature is indeed affected by this geological activity through how much ice is formed on the Great Lakes and the speed of the north magnetic dip pole's motion.

Arthur Viterito calculated the heat generated from this seismic activity to be 62% of the total heat causing global warming[37]. That means only 38% of the heat warming the earth is caused by other factors such as carbon dioxide and sun output variance.

NOAA explains the current debate as:

"The challenge in pinning down the exact relationship between a given carbon dioxide level and a specific temperature is that other factors influence temperature, too, including atmospheric dust and volcanic eruptions, amount of forests and other vegetation, and variations in incoming sunlight.

. . . In fact, the likely range of 2.7°- 8.1°F published in the recently released 5th Assessment report from the Intergovernmental Panel on Climate Change is the same as in their first report, published in 1990.

. . . climate experts think it is extremely unlikely that equilibrium warming from doubled carbon dioxide will be less than 2°F, and very unlikely that it will be more than 11°F."[41]

And since they were unaware of the geological heat being released and therefore were assuming the heat sources were only the carbon dioxide and sun variability, the actual result of doubling of carbon dioxide will probably be no less than 0.4°C (0.64°F) and no more than 2°C (3.52°F) for an average of 1.2°C (2.2°F) or less.

Chapter 16

What are we Looking at?

Doubling the carbon dioxide from pre-industrial values would take us to approximately 580ppm carbon dioxide for a temperature increase of about 1.8°.

We know that raising carbon dioxide in the atmosphere warms the earth but is not the major driver in the temperature of the earth. The earth's past has shown carbon dioxide itself to never have been the major player in earth's temperature, but more of a modifier to other major drivers. Because of the long-term history discussed above we know we have an exceptionally good handle on the overall effect of carbon dioxide

A warmer climate will kill fewer people and animals as seen in the Medieval Warm Period and the Roman Warm Period. During these times the earth's population went from gaining 13 million people per 100 years for 300 years to gaining 30 to 59 million people per 100 years for 400 years[8].

The earth is heading into a low insolation phase and its orbits around the sun are bringing it awfully close to the theoretical tripping point to begin a major glaciation. Since 560ppm carbon dioxide is well below the 5,000ppm max for vigorous human activity higher carbon dioxide ppm should not be a detriment.

Here is some of what we know at this point:

- Burning fossil fuels does not significantly affect the percentage of oxygen in the air and will not harm animals or humans.
- The present rate of temperature rise is no different than has occurred in the last 10,000 years when man was not an influence[16].
- What little effect humans had has reduced the Atlantic hurricane total energy[42].
- The total number of F3+ tornadoes has been on the decline since the mid 1960's[43].
- The total number of tornadoes is the same now as in 1954. Their intensity has dropped.
- Our drought index for the lower 48 states shows no long-term drought changes[44].

The previous statements indicate the present trend has been to calm the weather, contrary to people's fears that it would make the weather worse.

On a non-measured metric I'm sure most of you are familiar with, the weather in general seems to be a bit more compatible with you, your pets, and your livestock.

In the past, carbon dioxide spikes have helped stabilize the earth's climate to levels living beings could live within.

The 560ppm point will not get us above the 1,000ppm point, thus we won't melt Greenland and Antarctica.

And finally, if the temperature drops because of a decreasing solar output and decreasing geological activity, we could go from our present vegetation map, Exhibit A[45], to a map similar to the last glacial maximum, Exhibit B[45].

In Exhibit B, the red tinted areas are glaciers, ice deserts, or permafrost and the blue tinted areas are reduced vegetation due to seasonal freezing[46]. This scenario would most certainly cause extreme problems for mankind.

Exhibit A

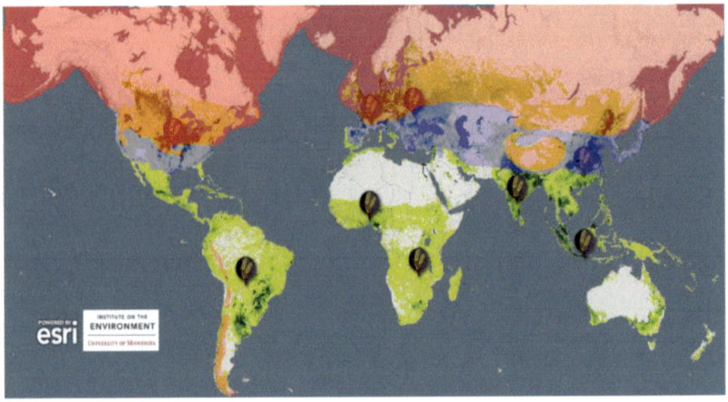

Exhibit B

Possible Occurrences and Dangers?

We have discussed many things which may affect the climate; some more significantly than others and some with higher risk of detrimental outcomes. Let's now talk more about things which may happen that could cause changes to the climate. I will use letters and Roman numerals to be able to discuss the list. Take a look:

A. Geological activity subsides (we have no control over this).
B. We stop burning carbon-based fuels (our choice).
C. Geological activity continues (we have no control over this).
D. We continue burning carbon-based fuels (our choice).

E. Solar output is steady (we have no control over this).
F. Solar activity drops (we have no control over this).
G. Geological activity increases (we have no control over this).

It is interesting to note that of the six possibilities listed four of them we have no control over, yet they can do much more harm or good to the earth than mankind can.

The primary reactions to the above which can cause harm to the earth consist of:

I. Property loss as in flooding by sea level rise or coverage by glaciers.
II. Property damage by weather events.
III. Ocean chemistry becoming acidified/alkaline.
IV. Humans having shorter life spans/more deaths per year.
V. Animals having shorter life spans/more deaths per year.
VI. Colder temps or less carbon dioxide will stress plants and the opposite will enhance them.

As you can see the damage is not just all or none but rather depends on what damage you are talking about. And the choices we have are not just burn carbon-based fuels or not burn carbon-based fuels. There are two natural forces that we have to contend with that can produce effects canceling out either the burn or not burn choice. Based on this we can say that the Pascal's Wager method is of no use to us and we need to analyze the situation based on what our reality is.

Chapter 17

What are our options?

In Chapter 16 we listed actions that can lead to climate change. In order to determine the best course of action, one must consider the four conditions: A&B, A&D, C&B, and C&D while E, F, or G happens.

I'm going to rate these activities based on what is probably going to happen in the next 2000 years or so. The very long term, 500,000 years, results are also dependent on these factors plus some we are not looking at. Thus, we cannot rate how they will affect that long of a time period.

To help determine what is the best option we will assign the following values to each problem based on a scale of 0 to 10:

> I Value of 2.50 - The loss of beach front property is of almost no concern to anyone, except, of course, those who own beach front property. It was still given a 2.50 value because there is some low-lying ground at or

not much above three feet above high tide level that is used for agriculture.

II Value of 5 - Again – property values are not that critical. People can learn to build properly for their hazards, but there will always be people who do not have the time to or just refuse to evacuate in areas that get hit by storms.

III Value of 7.5 - Aquatic life forms are a critical part of the food chain.

IV Value of 10 - Protecting human life should always be the top priority. This is a short-term consideration that only helps humans while the activity is present. Burning fuel only enhances the decay period of the carbon dioxide in the atmosphere.

V Value of 8.75 - We need animals for food, and carbon dioxide generators for the plants. We can't do it ourselves. This consideration has the same limitations as IV.

VI Value of 7.5 – Something to shoot for, but the fact is, as we advance, we will probably come up with different solutions.

If the condition moves in a good direction, a positive value is used, and if it moves us in a bad direction a negative value is used. If there is no way to determine the effect, then a value of 0 is used.

In the following section the text highlighted in blue represents situations that are best for mankind and those in yellow are not so good for mankind. Those with no highlighting are unchanged or unknown.

Analysis of each individual consideration.

A. The earth's geological activity subsides (we have no control over this).
 I. Property loss. Minor decrease in sea level.
 II. Property damage. Storms stay the same.
 III. Ocean chemistry. Unchanged, aquatic life is not harmed.
 IV. Humans. Shorter lifespan for all humans.
 V. Animals. Shorter lifespan for all animals.
 VI. Plants. More stress on plants as the earth's temperature drops slightly, loss of carbon dioxide & heat to oceans.

B. Stop burning carbon-based fuels (our choice).
 I. Property loss. Minor decrease in sea level.
 II. Property damage. Storms stay the same.
 III. Ocean chemistry. Unchanged, aquatic life is not harmed.
 IV. Humans. Shorter lifespan for all humans.
 V. Animals. Shorter lifespan for all animals.
 VI. Plants. More stress on plants as the earth's temperature drops slightly. Loss of carbon dioxide & heat to oceans and land.

C. The earths geological activity continues (we have no control over this).

I. Property loss. Minor rise of sea level.
II. Property damage. Storms stay the same.
III. Ocean chemistry. Changed by geological activity, aquatic life harmed.
IV. Humans. Human lifespan enhanced by warmth
V. Animals. Animal lifespan enhanced by warmth
VI. Plants. Plant life enhanced by carbon dioxide and warmth.

D. Continue burning carbon-based fuels (our choice).

I. Property loss. The planet warms and the seas rise 1 to 3 feet.
II. Property damage. Storms become stronger/storms do not change. Conflict in data and no possible solution.
III. Ocean chemistry. Changed by carbon dioxide, aquatic life harmed.
IV. Humans. Human life enhanced.
V. Animals. Animal life enhanced.
VI. Plants. Plant life enhanced by carbon dioxide and warmth.

E. Solar output steady (we have no control over this).

I. Property loss. No change in sea level.
II. Property damage. Storms stay the same.
III. Ocean chemistry. Aquatic life not harmed.

IV. Humans. Shorter lifespan for all humans.
V. Animals. Shorter lifespan for all animals.
VI. Plants. No change.

F. Solar activity drops (we have no control over this).
I. Property loss. Loss of significant farm hectares due to cold.
II. Property damage unknown.
III. Ocean chemistry, no change.
IV. Humans. Shorter lifespan per year due to cold.
V. Animals. Shorter lifespan per year due to cold.
VI. Plants. More stress due to cold.

G. Geological activity increases (we have no control over this).
I. Property loss. The planet warms and the seas rise 1 to 3 feet.
II. Property damage. Storms become stronger/storms do not change. Conflict in data and no possible solution.
III. Ocean chemistry. Aquatic life harmed.
IV. Humans. Lifespan increases due to warmth.
V. Animals. Lifespan increases due to warmth & enhanced plants.
VI. Plants. Enhanced through warmth and more carbon dioxide.

Looking back at this list we must remember that A to G are if that condition and that condition only is occurring. This is an artificial consideration because in reality we have all three conditions (geological activity, solar activity changes, and we are burning carbon based fuels) physically occurring now.

Assigning values to the above and giving them equal weight shows:

Because no one condition will ever change on its own but rather all three will be changing in one way or another, thus we can put the numbers from Table 1 into each condition.

Depending on how the numbers are analyzed, I can say that in the burn no fossil fuel and burn fossil fuel scenarios the results are:

Burn no fossil fuels average is - 3.00

Burn fossil fuels average is + 0.24

This indicates that in overall probability burning fossil fuels has the highest chance of being a positive result for mankind. On average, it gives us a roughly 2% better environment for mankind. The choice to not burn fossil fuels gives us a roughly 30% worse environment for mankind.

The same scenario using the number of positive outcomes instead of percentages would look like this:

Burn no fossil fuels positive outcomes are 0

Burn fossil fuels positive outcomes are 4

Individual Condition Stand Alone Results

Condition	I	II	III	IV	V	VI	Base 10 Result	
Importance	2.5	5	7.50	10.00	8.75	7.5		What Happens
A	2.50	0.00	0.00	-10.00	-8.75	-7.50	-5.76	Geo ▼
B	2.50	0.00	0.00	-10.00	-8.75	-7.50	-5.76	No fossil fuel
C	-2.50	0.00	-7.50	10.00	8.75	7.50	3.94	Geo Same
D	-2.50	0.00	-7.50	10.00	8.75	7.50	3.94	Use fossil fuel
E	2.50	5.00	7.50	-10.00	-8.75	0.00	-0.91	Solar Activity Unchanged
F	-2.50	0.00	0.00	-10.00	-8.75	-7.50	-6.97	Solar Activity Drops Off
G	-2.50	0.00	-7.50	10.00	8.75	7.50	3.94	Geo ▲

Table 1

Condition	I	II	III	IV	V	VI	Base 10 Result	What Happens	
Solar Activity Unchanged									
A & B	2.50	1.67	2.50	-10.00	-8.75	-5.00	-4.14	Geo ▼	No fossil fuel
C & B	0.83	1.67	0.00	-3.33	-2.92	0.00	-0.91	Geo Same	No fossil fuel
A & D	0.83	1.67	0.00	-3.33	-2.92	0.00	-0.91	Geo ▼	Use fossil fuel
C & D	-0.83	1.67	-2.50	3.33	2.92	5.00	2.32	Geo Same	Use fossil fuel
Solar Activity Drops Off									
A & B	0.83	0.00	0.00	-10.00	-8.75	-7.50	-6.16	Geo ▼	No fossil fuel
C & B	-0.83	0.00	-2.50	-3.33	-2.92	-2.50	-2.93	Geo Same	No fossil fuel
A & D	-0.83	0.00	-2.50	-3.33	-2.92	-2.50	-2.93	Geo ▼	Use fossil fuel
C & D	-2.50	0.00	-5.00	3.33	2.92	2.50	0.30	Geo Same	Use fossil fuel
Solar Activity Unchanged									
G & B	0.83	1.67	0.00	-3.33	-2.92	0.00	-0.91	Geo ▲	No fossil fuel
G & D	-0.83	1.67	-2.50	3.33	2.92	5.00	2.32	Geo ▲	Use fossil fuel
Solar Activity Drops Off									
G & B	-0.83	0.00	-2.50	-3.33	-2.92	-2.50	-2.93	Geo ▲	No fossil fuel
G & D	-2.50	0.00	-5.00	3.33	2.92	2.50	0.30	Geo ▲	Use fossil fuel

Table 2

This indicates that out of the 12 possible scenarios zero of them (0%) give a positive outcome for not burning fossil fuels and four of them (33%) give a positive outcome for burning fossil fuels.

That still leaves seven outcomes (59%) that will not be advantageous to mankind. We may or may not have an opportunity to change our use of fossil fuels to get a positive outcome. This is primarily due to the fact that there are two sources of heat for the earth both of which we have no control over. These are the variance of the sun and geothermal activity.

Williams determined with an 80% coefficient of determination that for temperature changes in the very near time period of 1880 to 1975 geothermal activity accounted for 51% of the change, carbon dioxide accounted for 30%, and 19% was from solar changes[47]. The question is, does that make a significant difference?

If we use those percentages instead of an equal weighting, we get:

Condition	I	II	III	IV	V	VI	Base 10 Result	What Happens	
Solar Activity Unchanged									
A & B	2.50	0.95	1.43	-10.00	-8.75	-6.08	-4.84	Geo ▼	No fossil fuel
C & B	-0.05	0.95	-2.40	0.20	0.18	1.58	0.11	Geo Same	No fossil fuel
A & D	1.00	0.95	-0.83	-4.00	-3.50	-1.58	-1.93	Geo ▼	Use fossil fuel
C & D	-1.55	0.95	-4.65	6.20	5.43	6.08	3.02	Geo Same	Use fossil fuel
Solar Activity Drops Off									
A & B	1.55	0.00	0.00	-10.00	-8.75	-7.50	-5.99	Geo ▼	No fossil fuel
C & B	-1.00	0.00	-3.83	0.20	0.18	0.15	-1.04	Geo Same	No fossil fuel
A & D	0.05	0.00	-2.25	-4.00	-3.50	-3.00	-3.08	Geo ▼	Use fossil fuel
C & D	-2.50	0.00	-6.08	6.20	5.43	4.65	1.87	Geo Same	Use fossil fuel
Solar Activity Unchanged									
G & B	-0.05	0.95	-2.40	0.20	0.18	1.58	0.11	Geo ▲	No fossil fuel
G & D	-1.55	0.95	-4.65	6.20	5.43	6.08	3.02	Geo ▲	Use fossil fuel
Solar Activity Drops Off									
G & B	-1.00	0.00	-3.83	0.20	0.18	0.15	-1.04	Geo ▲	No fossil fuel
G & D	-2.50	0.00	-6.08	6.20	5.43	4.65	1.87	Geo ▲	Use fossil fuel

Table 3

Again, depending on how these numbers are analyzed, I can say that the average of burn no fossil fuel and burn fossil fuel scenario outputs being good is:

Burn no fossil fuels average is - 2.12

Burn fossil fuels average is + 0.79

For not burning fossil fuels there are 2 positive outcomes with a value of 0.11 each so the average will also be 0.11 or 1.1%.

This indicates that in overall probability burning fossil fuels has the highest chance of being a positive result for mankind. On average it gives a roughly 8% better environment for mankind. The choice to not burn fossil fuels gives a roughly 21% worse environment for mankind.

The same scenario using the number of positive outcomes instead of percentages would look like this:

Bun no fossil fuels positive outcomes are 2

Burn fossil fuels positive outcomes are 4

This indicates that out of the 12 possible scenarios two (17%) give us a better environment for mankind if we stop burning fossil fuels and four (33%) give us a better environment for mankind if we continue burning fossil fuels.

In the equally weighted case, we are left with seven outcomes (59%) that will not be advantageous to mankind. In the weighted by heat addition percentage case we are left with six outcomes (50%) that regardless of what we do the outcome is negative. Again, in both cases this is because there are two sources of heat for the earth both of which we have no control over.

Combining the results from both weighting methods, it is most probable that there is a 46% chance of making the right choice.

Since we must make a choice, and averaging the two methods burning fossil fuels gives a 33% chance of getting a 5% better environment and not burning fossil fuels gives an 8% chance of getting a 26% worse environment.

Since there are so many numbers a quick review shows that:

Balanced or Heat Percent	Burn or No Burn	Chance Burn/ No burn correct	Environment % Change		% Chance we can't control
Balanced	Burn	33%	2%	Better	59%
Balanced	No Burn	0%	30%	Worse	59%
Heat Percent	Burn	33%	8%	Better	50%
Heat Percent	No Burn	17%	1%	Better	50%
Average of both	Burn	33%	5%	Better	55%
	No Burn	8%	15%	Worse	55%

Table 4

The "% Chance we can't control what happens" column is the probability that either the Suns activity or geothermal activity or both will work against us and we will not be able to control the resulting cold or warm change.

When considering all variables, the conclusion is: It is more favorable to burn carbon-based fuels and to stay below 1000ppm carbon dioxide.

Our second best option is to assume that solar activity will be unchanged and that geological activity change will do minimum degradation to the environment, thus the choice to not burn fossil fuels is viable. This places hope that the laws of physics will turn out to be what we want them to be. Since the average of the two No Burn scenarios is 8% that implies there is an 8% chance that it will work out that way.

The worst combination is solar activity dropping, geological activity slowing down, and no longer burning carbon-based fuels. With earth's orbit around the sun being the way it is now this scenario could bring another major glacial era. Something that would quite literally destroy our present technologically based societies.

There is an argument for looking at only the short-term (as in the next 20 to 50 years) effects of burning carbon-based fuels and relying on technological advances to address other problems that may arise in the next several generations. The problem with this argument is we don't have the capability to develop technologies to overcome change in the output of the sun or to change the way heat is released from the core of the earth, both of which are well beyond our present knowledge, let alone technologies.

There are also three additional conditions to keep in mind. Item VI, colder temperatures and less carbon dioxide, will probably be addressed in the future as we learn more and our agricultural technology advances. In addition, Item II, property damage by weather, is presently disputed depending on which government data/opinions you use. But even under worst case conditions, the earth and life in general has survived robustly when both the carbon dioxide and temperature was much higher. Contrary to popular belief, life has not done very well with decreasing/colder temperatures.

The final consideration is whether to consider life in general less important than the short-term gain in wealth. In that case this analysis will do no good because it would place too much emphasis on life and not enough emphasis on money.

Contrary to the present perception of the public, there is no easy answer to how to respond to climate change. Simply adding carbon dioxide to the atmosphere is not the single cause of climate change that has been occurring. The situation is not the same as when physics was deriving single equations that described how things work, such as $E = mc^2$ in which you measure the resultant mass and you know c. Instead we are working with many different variables causing effects and many different things that are being affected.

My own opinion is to take the safe route and continue burning carbon-based fuels until we can reach between 560ppm and 1000 ppm carbon dioxide and in the mean time invest in developing fusion technology and environmentally friendly technologies.

I hope this information will help you make a reasonable decision about what you want to subject all life on earth to.

Sources & Notes

1 Stanford Encyclopedia of Philosophy, Pascal's Wager https://plato.stanford.edu/entries/pascal-wager/

2 Berner & Kothavala (2001) GEOCARB III CO2 Scotese (2008) 'warm'/'cool' schematic

3 29 CFR 1910.1000 and BLM pub 25apxC.pdf

http://www.blm.gov/style/medialib/blm/wy/information/NEPA/cfodocs/howell.Par.2800.File.dat/25apxC.pdf

4 "Emergency and Continuous Exposure Guidance Levels for Selected Submarine Contaminants ", ISBN: 0-309-10661-3, *http://www.nap.edu/catalog/11170.html*

5 *http://hyperphysics.phy-astr.gsu.edu/hbase/Biology/phoc.html*

6 Laci M. Gerhart and Joy K. Ward, 2010 "Plant responses to low [CO2] of the past",New Phytologist (2010), *doi: 10.1111/j.1469-8137.2010.03441.x*

7 *http://www.cdc.gov/nchs/datawh/statab/unpubd/mortabs/gmwkIV_10.htm*

8 https://ourworldindata.org/grapher/absolute-increase-global-population?time=earliest..2029

9 Lucas, S.G., 2005, 25 years of mass extinctions and impacts: Geotimes, February, p. 28-32.

10 Courtillot, V., 1999, Evolutionary catastrophes -- the science of mass extinction: Cambridge, UK, Cambridge University Press, 173 p., ISBN 0-521-58392-6

11 Raup, D. M. and Sepkoski, J. J., Jr., 1982, Mass extinctions in the marine fossil record: Science, v. 215, p. 1501-1503.

12 Raup, D.M., and Sepkoski, J.J., Jr., 1984, Periodicity of Extinctions in the Geologic Past: Proceedings of the National Academy of Sciences (USA), v. 81, p. 801-805.

13 Mass Extinctions and Their Aftermath by A. Hallam , P. B. Wignall (1997) ISBN-13:978-0198549161,ISBN-10:0198549164

14 https://www.baylor.edu/geosciences/index.php?id=953452#extinctions

15 A Correlated History of Earth Poster 7th Edition, Paul R. Janke, Pan Terra Inc.

16 *Temperature Estimates – Ice Cores, Zachos 2001 dO18 Isotopes, Veizer dO18 Isotopes, CO2 Estimates – Geocarb III*

17 "Flood Basalt Volcanism During the Past 250 Million Years", Michael R. Rampino and Richard

B. Stothers, *Science 241 (4866): 663–668. Also available as a pdf at http://pubs.giss.nasa.gov/docs/1988/1988_Rampino_Stothers_1.pdf*

18 *www.cdc.gov/niosh/docket/archive/pdfs/NIOSH-125/125-AtmosphereManagementF-M.pdf*

19 "A New Model For Atmospheric Oxygen Over Phanerozoic Time", Robert A. Brener and Donald E. Canfield, American Journal Of Science, Vol 289, April 1989 P 333-361

20 Ralph Keeling, Scripps Institution of Oceanography (Scripps oxygen Program) 2014, *http://cdiac.ornl.gov/trends/oxygen/modern_records.html*, June 08, 2014

21 "Plant physiology", Taiz & Zeiger, 1991 ISBN 080530245X, 9780805302455

22 https://www.climate.gov/news-features/understanding-climate/climate-change-global-temperature, History of global surface temperature since 1880 interactive graph.

23 *http://usda.mannlib.cornell.edu/MannUsda/viewDocumentInfo.do?documentID=1860*

24 *ftp://aftp.cmdl.noaa.gov/products/trends/co2/ co2_annmean_mlo.txt*

25 *Plots from data at 23 above.*

26 *www.worldometers.info/world-population/*

27 World's farmland total bigger than expected https://www.producer.com/2017/11/worlds-farmland-total-bigger-expected/

28 About 67 miles by 67 miles every year. Which is about 232 miles by 232 miles for this 12 year period.

29 Richard Cowen, 1999, *http://www.ucmp.berkeley.edu/education/events/cowen1b.html*

30 Alley, R.B.. 2004.

GISP2 Ice Core Temperature and Accumulation Data.
IGBP PAGES/World Data Center for Paleoclimatology
Data Contribution Series #2004-013.
NOAA/NGDC Paleoclimatology Program, Boulder CO, USA.

31 Jason E. Box, Lei Yang, David H. Bromwich, and Le-Sheng Bai, Journal of Climate 2009 "Greenland Ice Sheet Surface Air Temperature Variability: 1840–2007", 22:14, 4029-4049 ,DOI: *http://dx.doi.org/10.1175/2009JCLI2816.1*

32 Paul Clark, 2016, *http://woodfortrees.org/plot/hadcrut4gl/from:1920/to:2016/mean:120/plot/hadcrut4gl/from:1975/to:2005/trend/plot/hadcrut4nh/from:1920/to:2016/mean:120/plot/hadcrut4nh/from:1975/to:2005/trend*

33 M. A. Kominz, "Sea Level Variations Over Geologic Time" Western Michigan University,

Kalamazoo, MI, USA Copyright © 2001 Academic Press doi:10.1006/rwos.2001.0255

34 Sara Raper, http://www.cru.uea.ac.uk/documents/421974/

1295957/Info+sheet+%2310.pdf/30611b7b-888d-45cb-bee3-1575b14e2e70

35 Dana L. Royer, 2005, "CO2-forced climate thresholds during the Phanerozoic", Science Direct, 2005.11.30

36 Archer & Ganapolski, 2005, "A movable trigger: Fossil fuel CO2 and the onset of the next glaciation", G3 Volume 6, Number 5. 2005.05.05 Q05003, *doi:10.1029/2004GC000891*, ISSN 1525-2017

37 "The Correlation of Seismic Activity and Recent Global Warming",Viterito, J Earth Sci Clim Change 2016, 7:4, *http://dx.doi.org/10.4172/2157-7617.1000345*

38 *"The Correlation of North Magnetic Dip Pole Motion and Seismic Activity" Williams, J Geol Geophys 2016, 5:6 http://dx.doi.org/10.4172/2381-8719.1000262*

39 *Paul Clark, 2016, http://woodfortrees.org/plot/hadcrut4gl/from:1920/to:2016/mean:120/plot/hadcrut4gl/from:1975/to:2005/trend/plot/hadcrut4nh/from:1920/to:2016/mean:120/plot/hadcrut4nh/from:1975/to:2005/trend*

40 *http://www.glerl.noaa.gov/data/ice/imgs/IceCoverAvg1973 _2015.jpg*

41 *https://www.climate.gov/news-features/climate-qa/how-much-will-earth-warm-if-carbon-dioxide-doubles-pre-industrial-levels*

42 Chris Landsea, 2015 National Hurricane Center, *http://www.aoml.noaa.gov/hrd/tcfaq/E11.html*

43 NCEI, 2016 *https://www.ncdc.noaa.gov/climate-information/extreme-events/us-tornado-climatology/trends*

44 *EPA, 2018 https://www.epa.gov/climate-indicators/climate-change-indicators-drought*

45 *Institute On The Environment, University of Minnesota http://environment.umn.edu/discovery/gli/collaborations/*

46 *K. Saito et el, 2013 "LGM permafrost distribution: how well can the latest PMIP multi-model ensembles perform reconstruction?", Clim. Past, 9, 1697–1714, doi:10.5194/cp-9-1697-2013*

47 Williams B (2020) The Oceans Surface Heat Sources Allocation. J Geol Geosci 4(1): 001-007 *http://sciaeon.org/ articles/The-Oceans-Surface-Heat-Sources%20Allocation.pdf*

Made in the USA
Las Vegas, NV
07 November 2026

57673975R00064